Teacher's Edition

PRECALCULUS

A Graphing Approach

Fourth Edition